MAKE 3 EQUAL PART

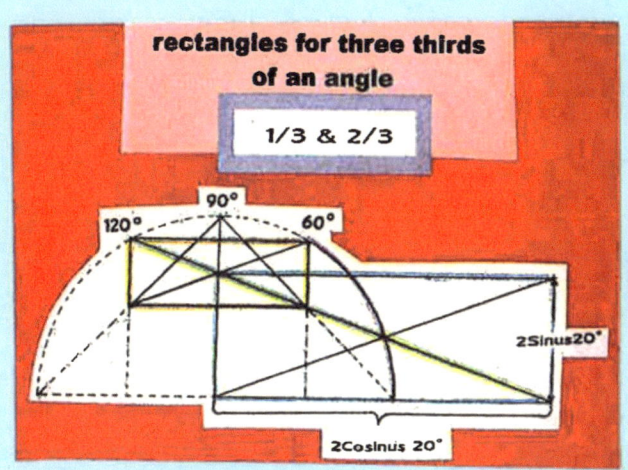

rectangles for three thirds of an angle

1/3 & 2/3

120° 90° 60°

2Sinus20°

2Cosinus 20°

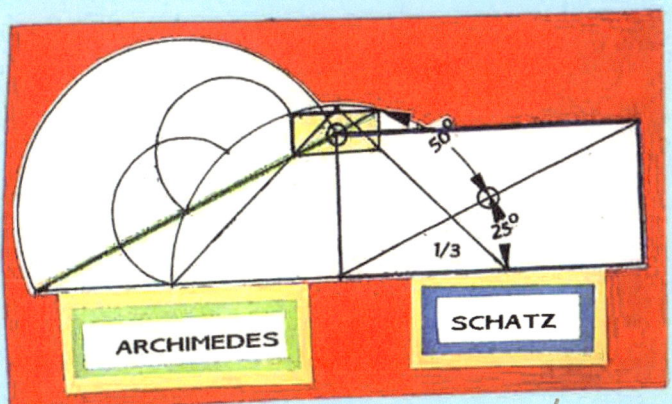

ARCHIMEDES

SCHATZ

50°
25°
1/3

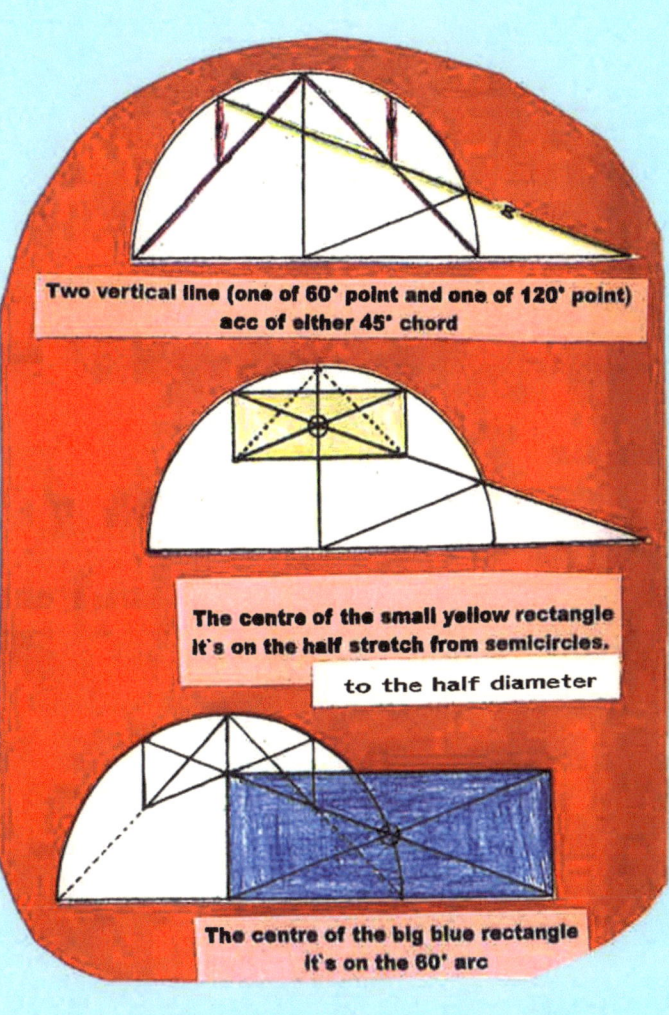

Two vertical line (one of 60° point and one of 120° point) acc of either 45° chord

The centre of the small yellow rectangle it's on the half stretch from semicircles.

to the half diameter

The centre of the big blue rectangle it's on the 60° arc

FROM WHAT ANGLE YOU LIKE

**obtuse angle
greater as a fourth
of circle**

**make with bisect
small as it**

*equal part
from arc=
equal angle*

© 2025 Harald Schatz
Verlag: BoD · Books on Demand GmbH,
Überseering 33, 22297 Hamburg,
bod@bod.de
Druck: Libri Plureos GmbH,
Friedensallee 273, 22763 Hamburg
ISBN: 978-3-8192-9818-9

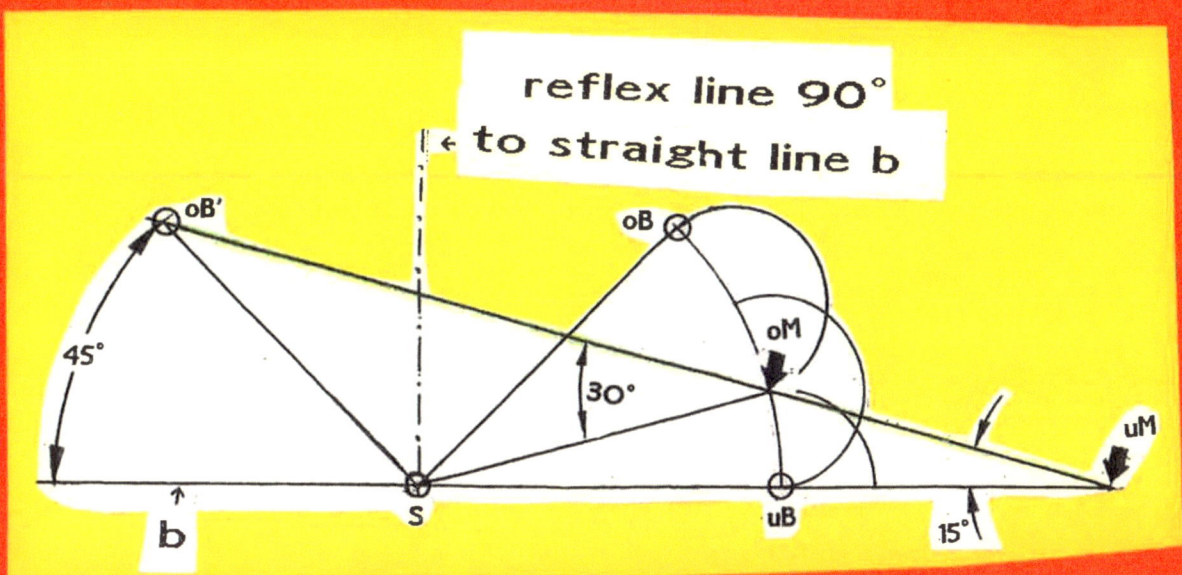

reflex line 90°
← to straight line b

the both diameter stretches are equal long

$$uB' - uB = Lg - uM$$

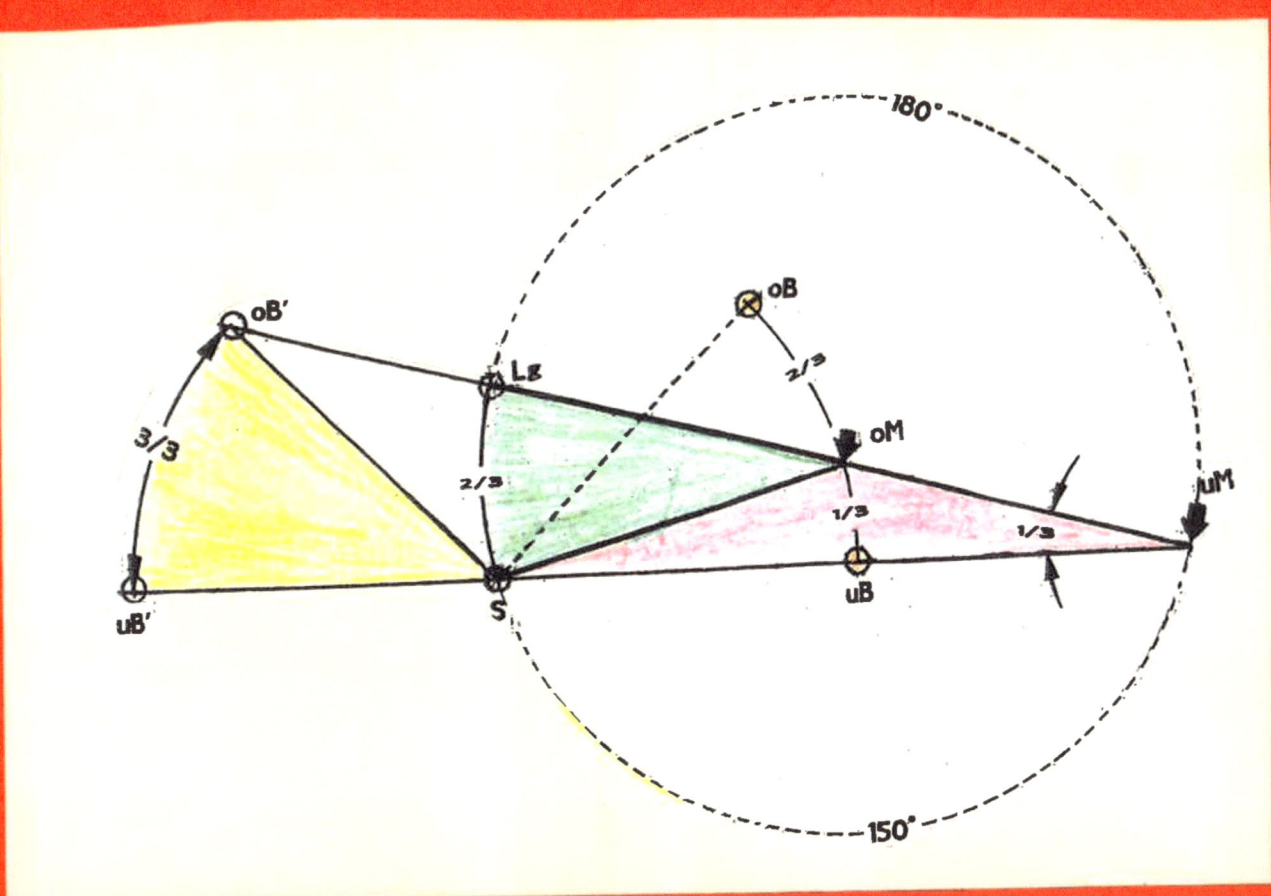

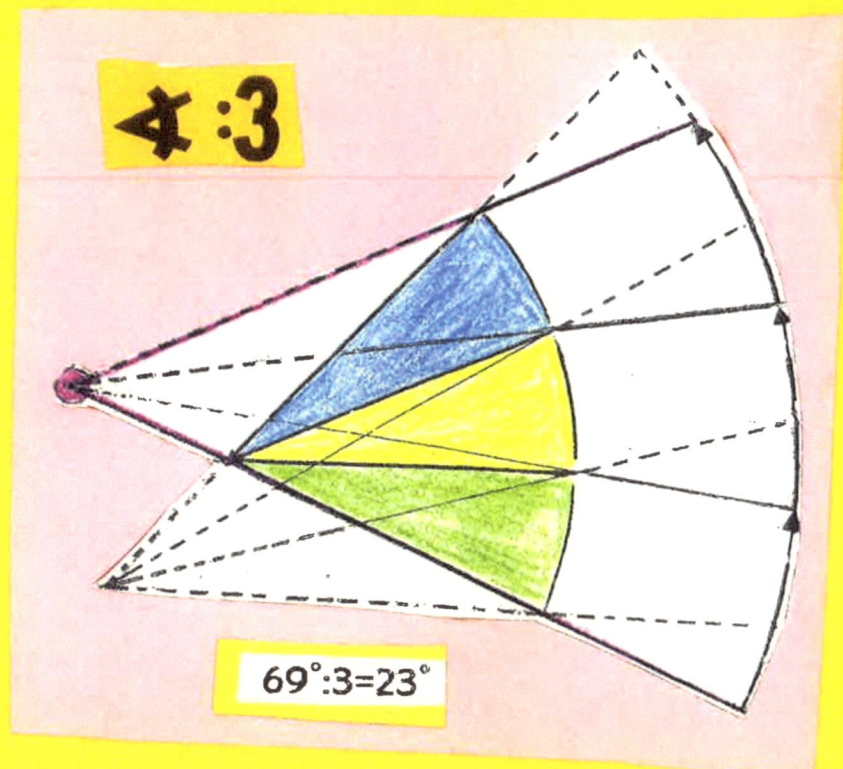

4:3

69/3

69°:3=23°

development

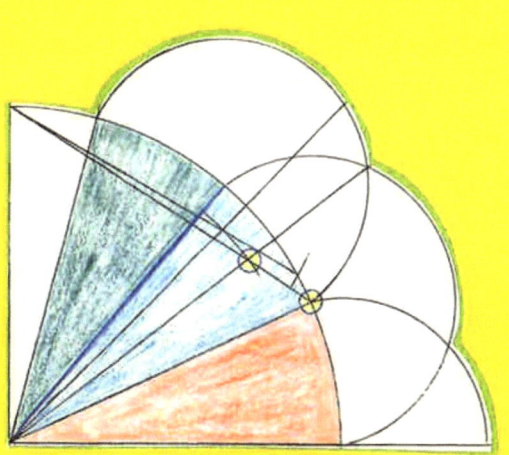

75/3

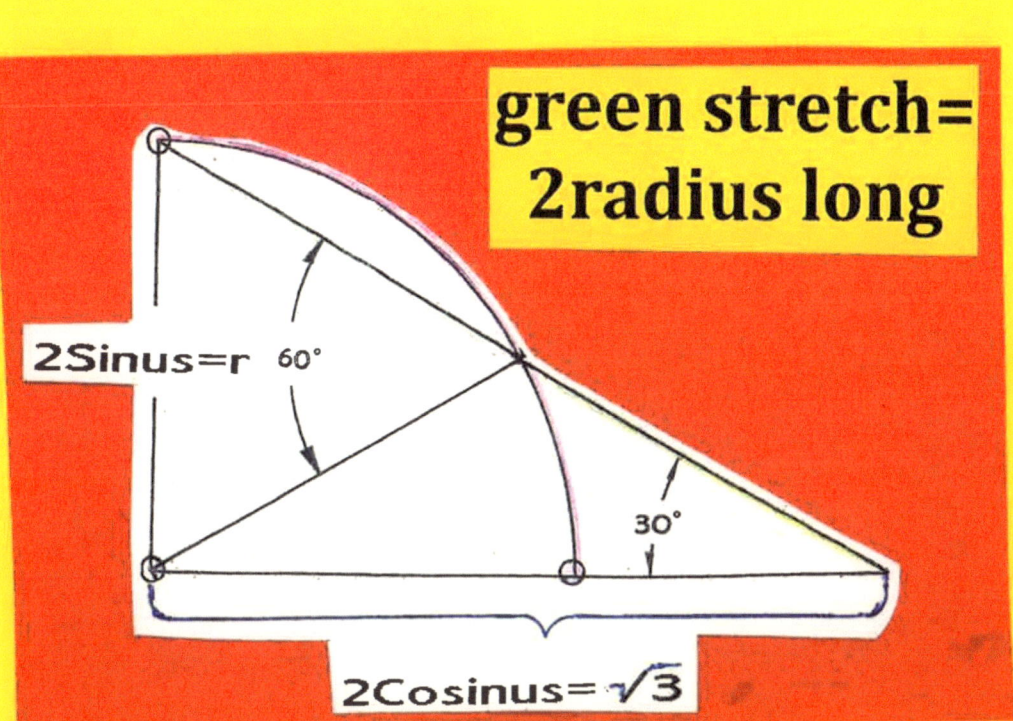

green stretch= 2radius long

90/3

2Sinus=r   60°

30°

2Cosinus=√3

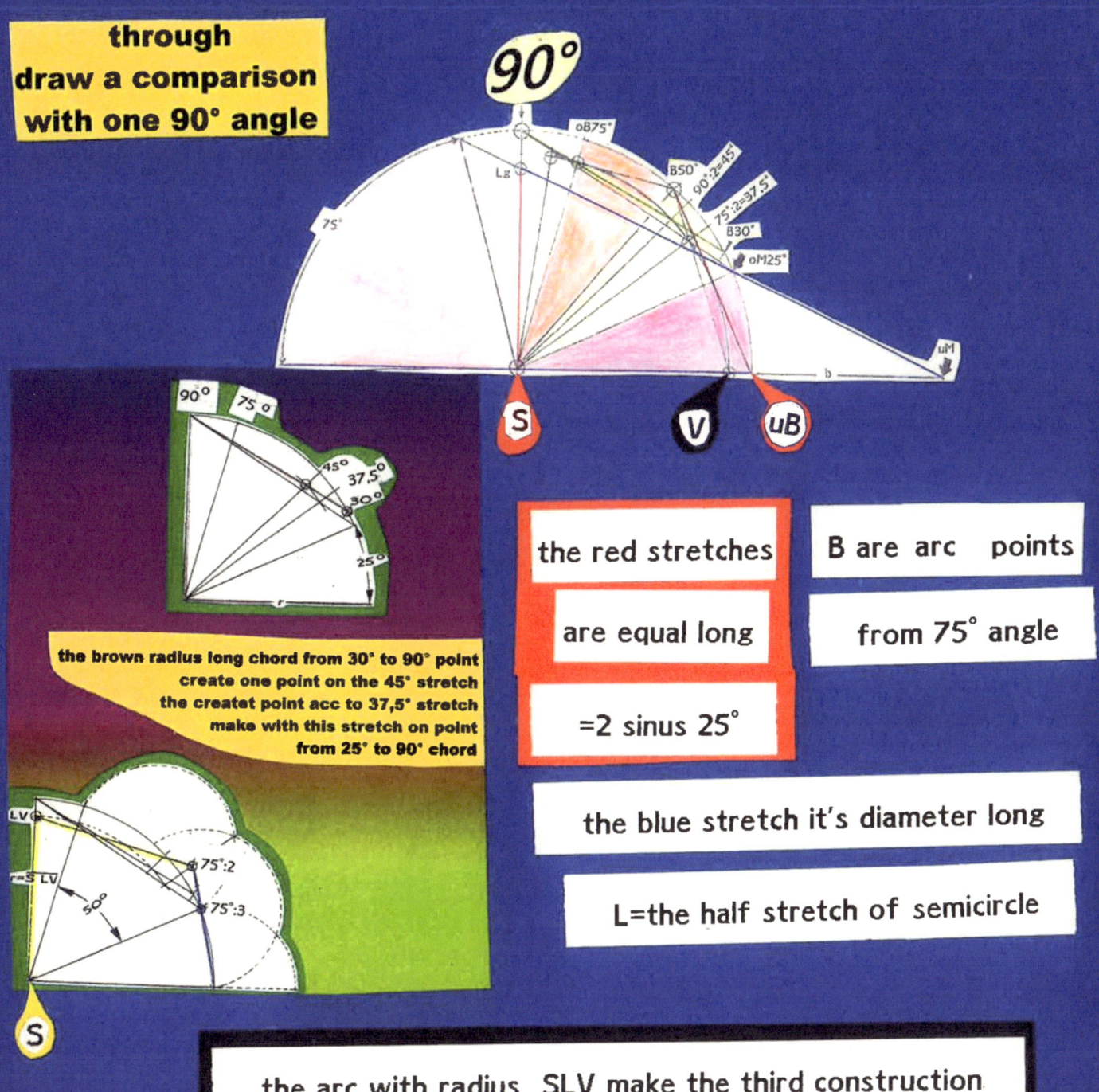

through
draw a comparison
with one 90° angle

90°

the brown radius long chord from 30° to 90° point
create one point on the 45° stretch
the createt point acc to 37,5° stretch
make with this stretch on point
from 25° to 90° chord

the red stretches

B are arc points

are equal long

from 75° angle

=2 sinus 25°

the blue stretch it's diameter long

L=the half stretch of semicircle

the arc with radius SLV make the third construction

three possible
methods to make a third of an angle

The impossibility

construction
of a third angle after
Harald Schatz rectangles method

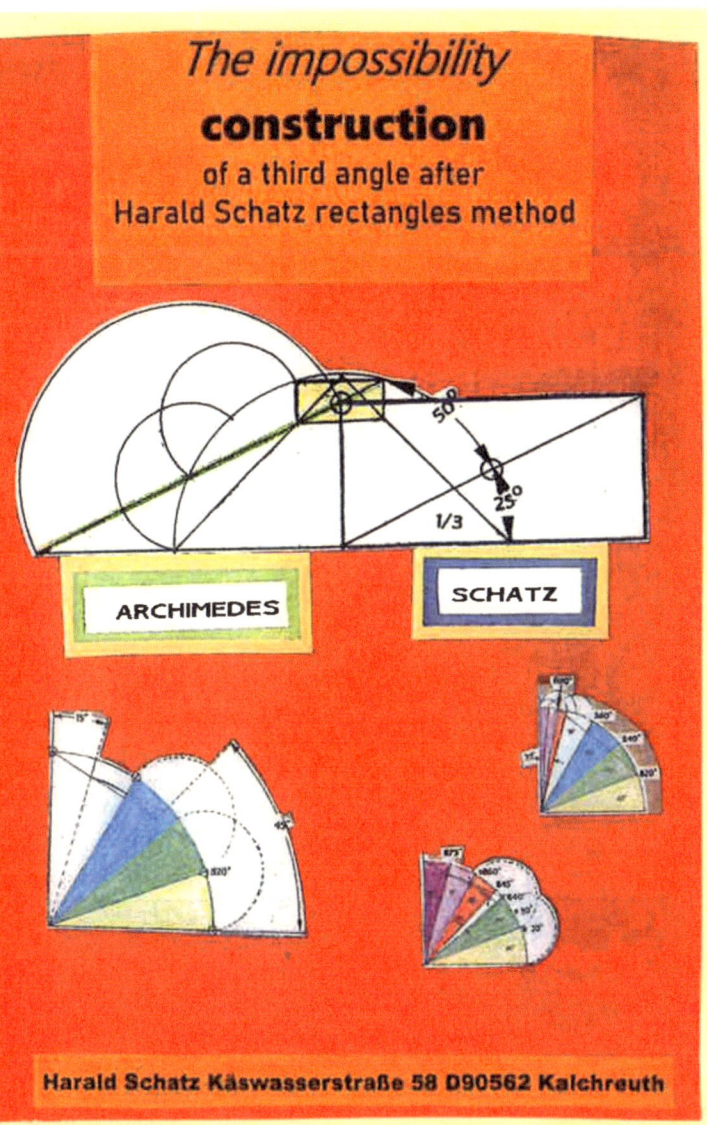

ARCHIMEDES

SCHATZ

Harald Schatz Käswasserstraße 58 D90562 Kalchreuth

a third angle construction with four
equal triangles under the straight line
to geometry proof after
Archimedes ruler method

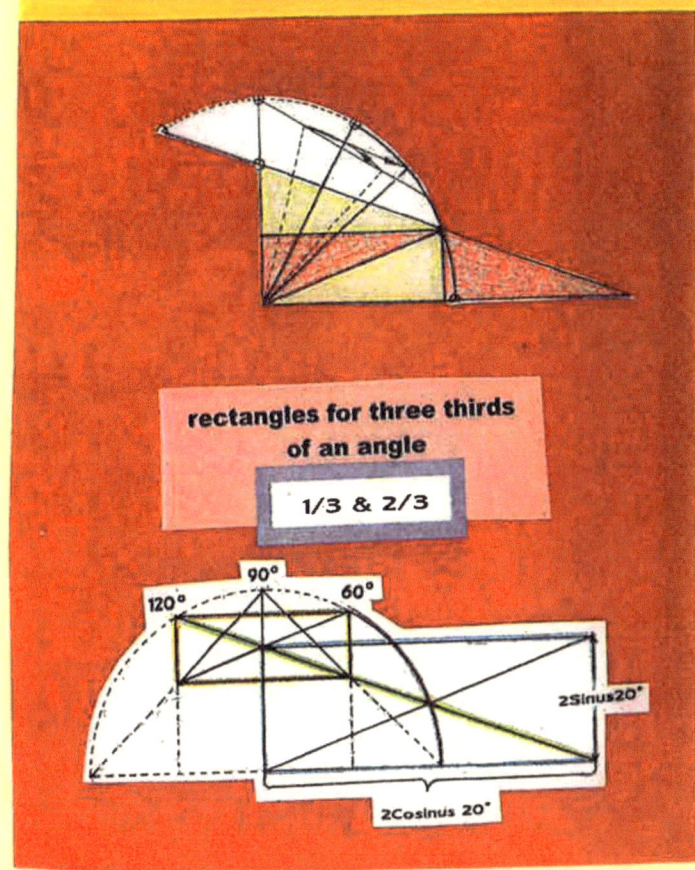

rectangles for three thirds
of an angle

1/3 & 2/3

How can you making a third
of one angle with two rectangles?

Two vertical line (one of 60° point and one of 120° point)
acc of either 45° chord

The centre of the small yellow rectangle
it`s on the half stretch from semicircles.

to the half diameter

The centre of the big blue rectangle
it`s on the 60° arc

The science
axiom from

Mister
Galois

it`s wrong!

three thirds of one angle

the first attempt
22.10.2004

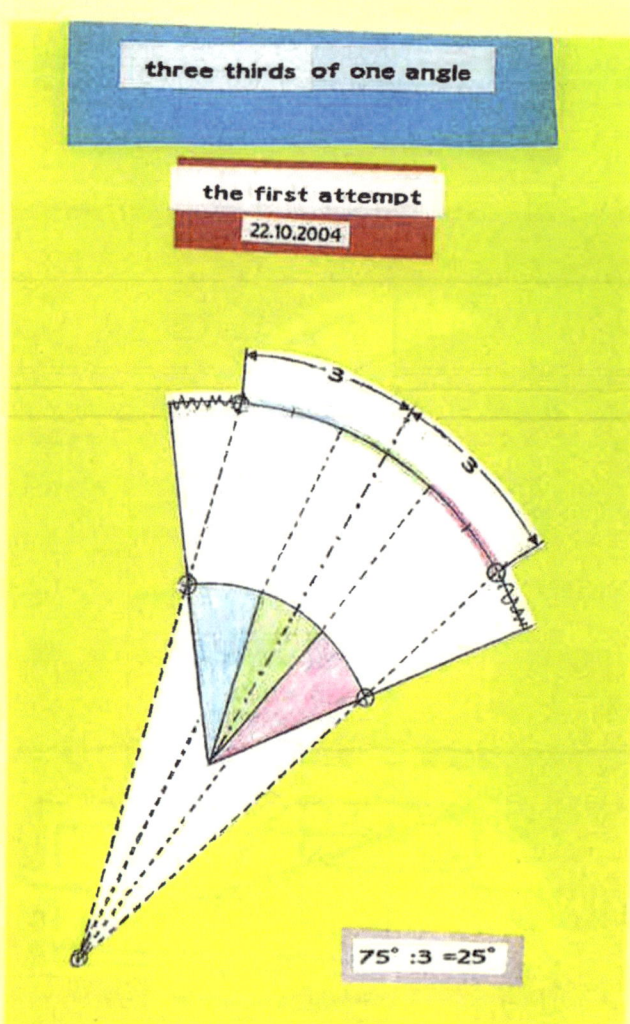

75° :3 =25°

from periphery angle

to centrical angle

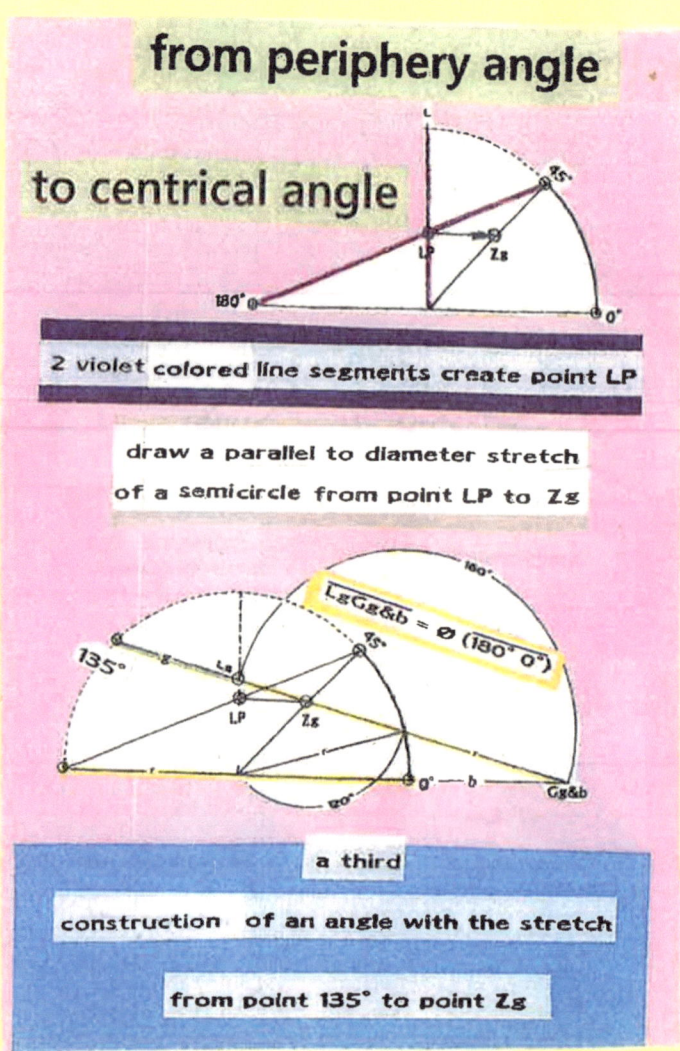

2 violet colored line segments create point LP

draw a parallel to diameter stretch
of a semicircle from point LP to Zg

LgGg&b = Ø (180° 0°)

a third

construction of an angle with the stretch

from point 135° to point Zg

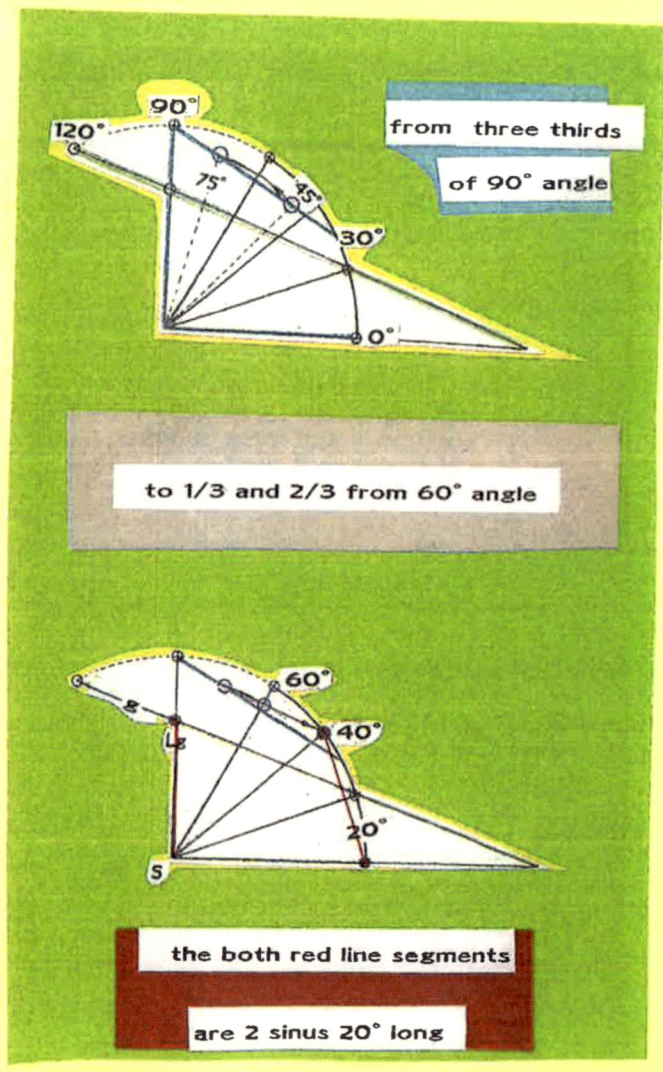

from three thirds
of 90° angle

to 1/3 and 2/3 from 60° angle

the both red line segments

are 2 sinus 20° long

Construction
of an 20° angle with
two semicircles

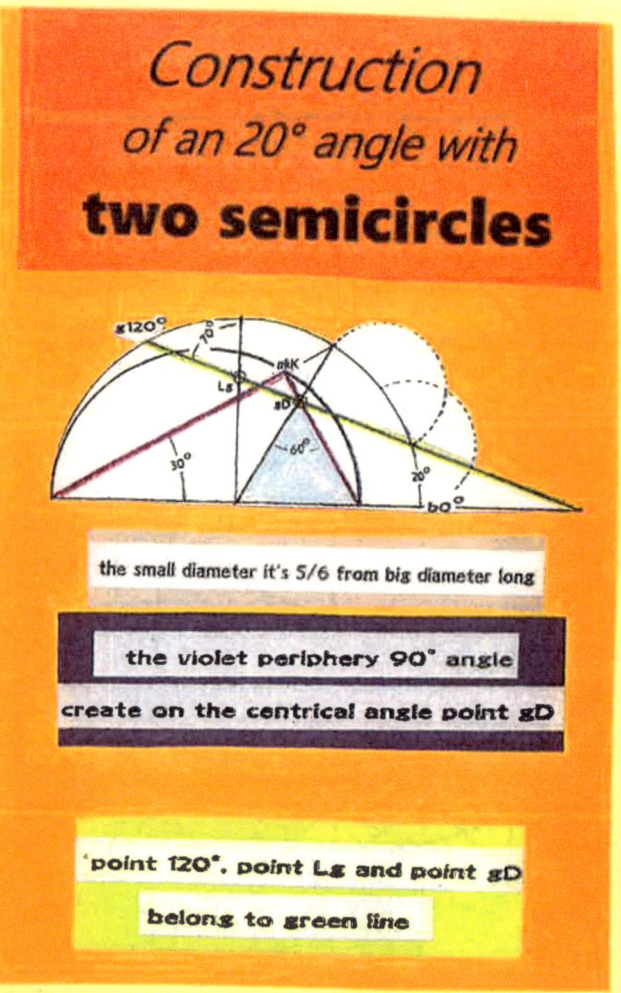

the small diameter it's 5/6 from big diameter long

the violet periphery 90° angle

create on the centrical angle point gD

point 120°, point Lg and point gD

belong to green line

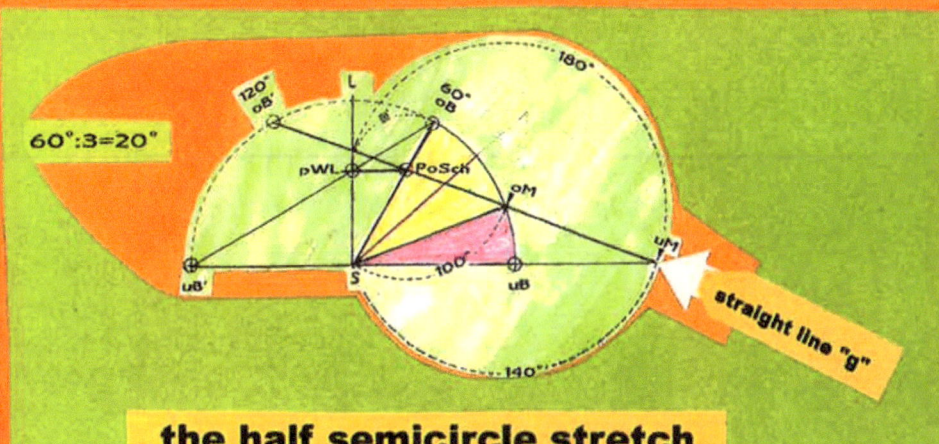

$60°:3=20°$

the half semicircle stretch to half diameter (uB`- uB) make with the periphery angle stretch point "pWL"
a Parallel to diameter from this point to centrical angle create point PoSch on the straight line g (proof of Archimedes)

three half part from the radius stretch and one equilateral triangle create with the green stretch on the 30°arc 20° and 10°

one and a half radius stretch
acc to three equal part of one angle

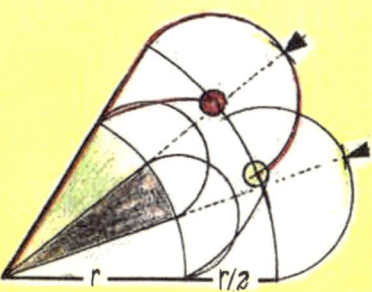

acc to with red and green arcs

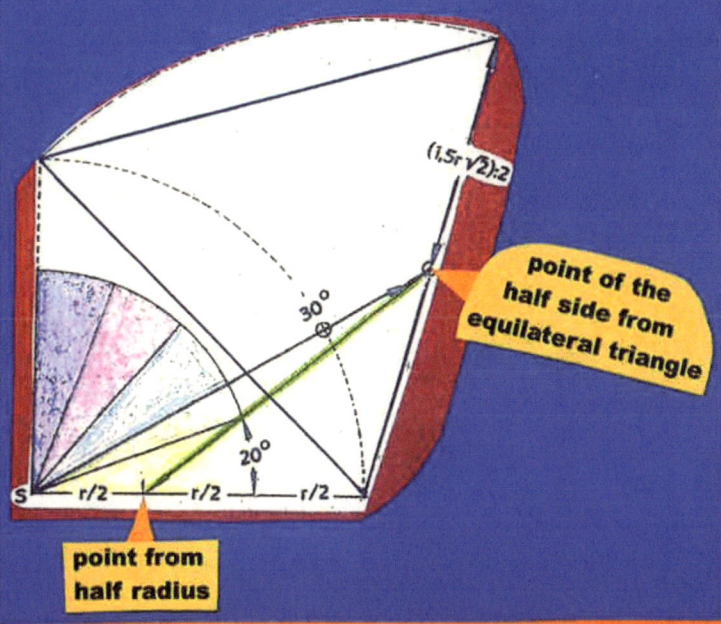

point of the half side from equilateral triangle

point from half radius

The impossibility construction

3/3

1/3

from a third of one 60° angle

only with a pair of compasses and ruler

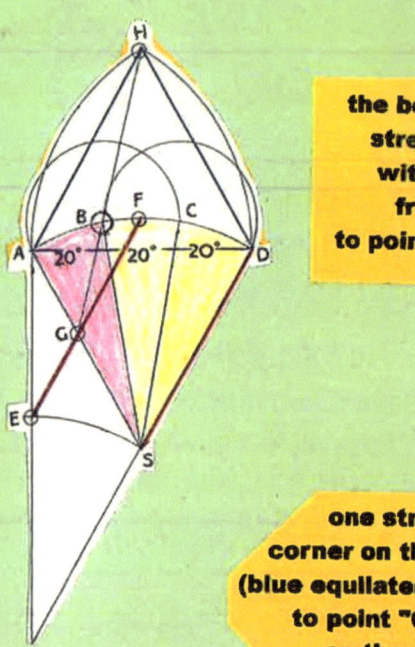

H

B F C
A 20° 20° 20° D
G
E
S

the both red parallel stretches create with the stretch from point "F" to point "E" point "G"

one stretch from corner on the point "H" (blue equilateral triangle) to point "G" create on the arc (A-D) point "B"

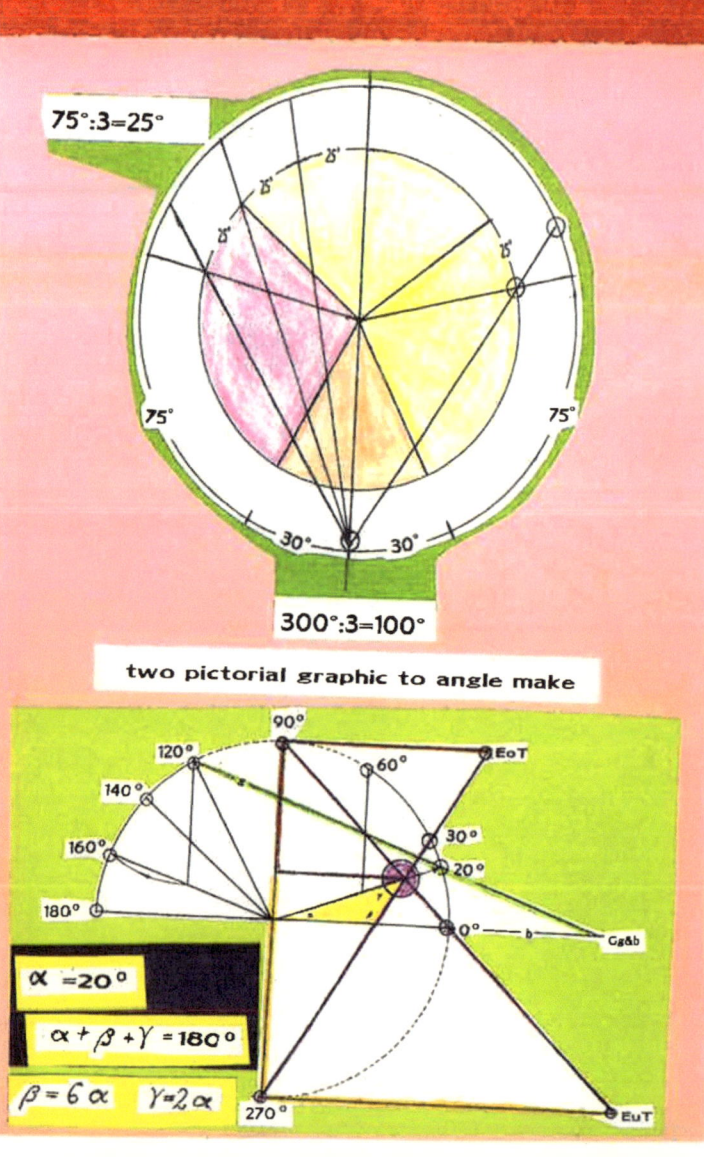

75°:3=25°

75°   75°

30°   30°

300°:3=100°

two pictorial graphic to angle make

90°
120°   60°   EoT
140°
160°   30°
20°
180°   0°ー b   Gg&b

α =20°

α + β + γ =180°

β = 6 α   γ=2 α

270°   EuT

from 2/3 and 1/3 diameter to 1/3 and 2/3 angle

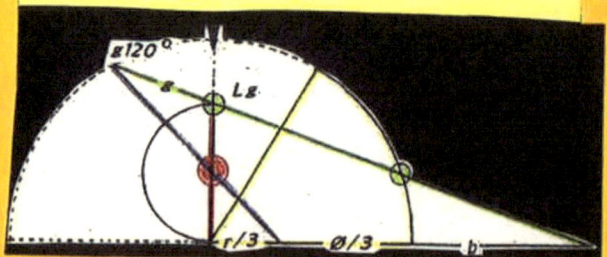

g 120
Lg
r/3   ø/3   b

the bisect stretch from the semicircle creat with the violet stretch the point red. Both red stretches from this starting point are sinus 20° long the starting point from green diameter stretch make with this stretch the divide from arc therefore from 60° angle

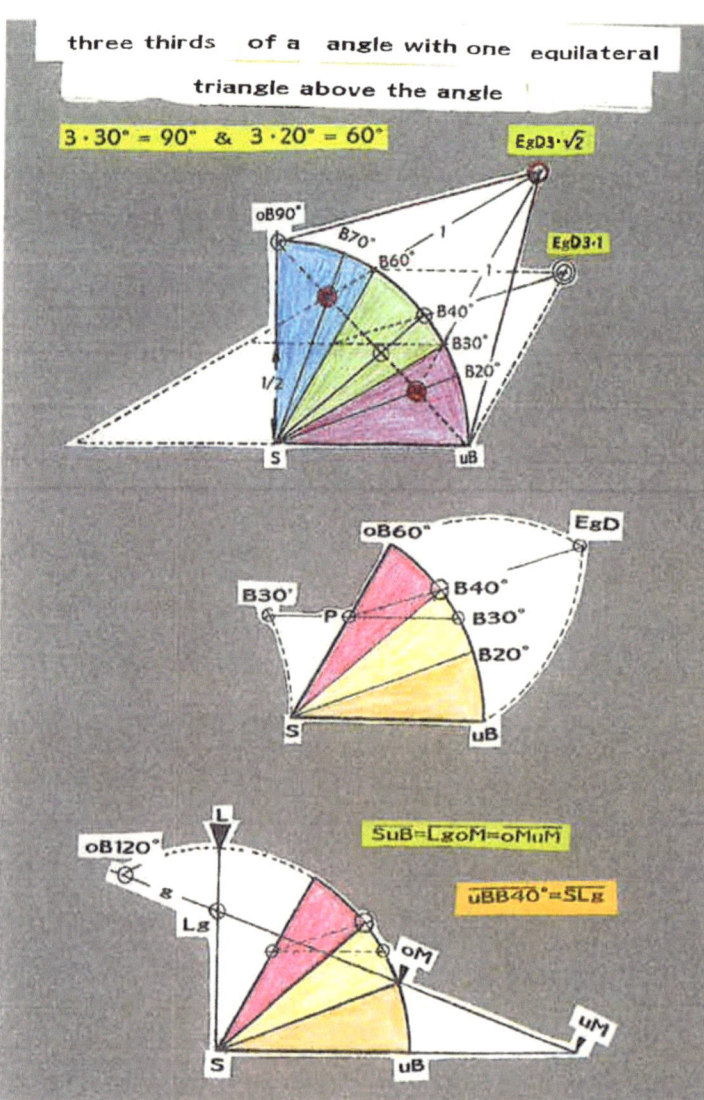

$3 \cdot 30° = 90°$ & $3 \cdot 20° = 60°$

EgD3·√2

oB90°   B70°
B60°
EgD3·1
B40°
B30°
B20°

1/2

S   uB

oB60°   EgD
B30°   B40°
P   B30°
B20°
S   uB

oB120°   L
g
Lg
oM
S   uB   uM

SuB=CgoM=oMuM

uBB40°=SLg

from three part
of secound arc

45°   60°

first   secound
arc   arc

first and
secound arc

5°
10°   15°
10°   15°
10°   15°

to three part
of first arc

a third of 75° angle
through
draw a comparison
with one 90° angle

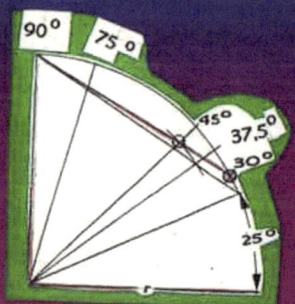

90°   75°
45°
37,5°
30°
25°
r

**the brown radius long chord from 30° to 90° point
create one point on the 45° stretch
the createt point acc to 37,5° stretch
make with this stretch on point
from 25° to 90° chord**

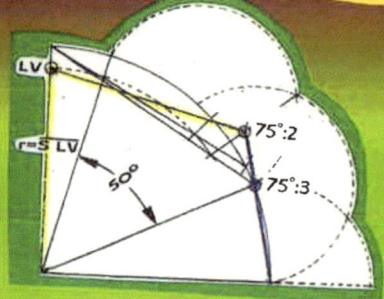

LV
r=S LV
50°
75°:2
75°:3

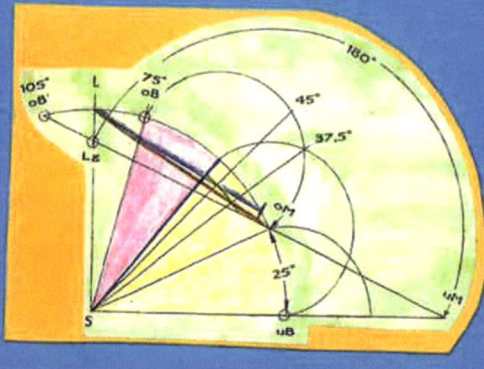

105°   L   75°   180°
oB   oB
45°
Lg   37,5°
oM
25°
S   uB   uM

draw a comparison
from 90°angle with 75°angle

make both angles with bisection comparable

to first on the 90°angle. the blue stretch create with
straight line from half angle on intersections point

make from this point one arc (with radius to corner "S")
to half straight line from 75° angle.

Make with the brown stretch one line from
90° point to point "oM" over the intersections point
on 37,5° streight to arc of both angles

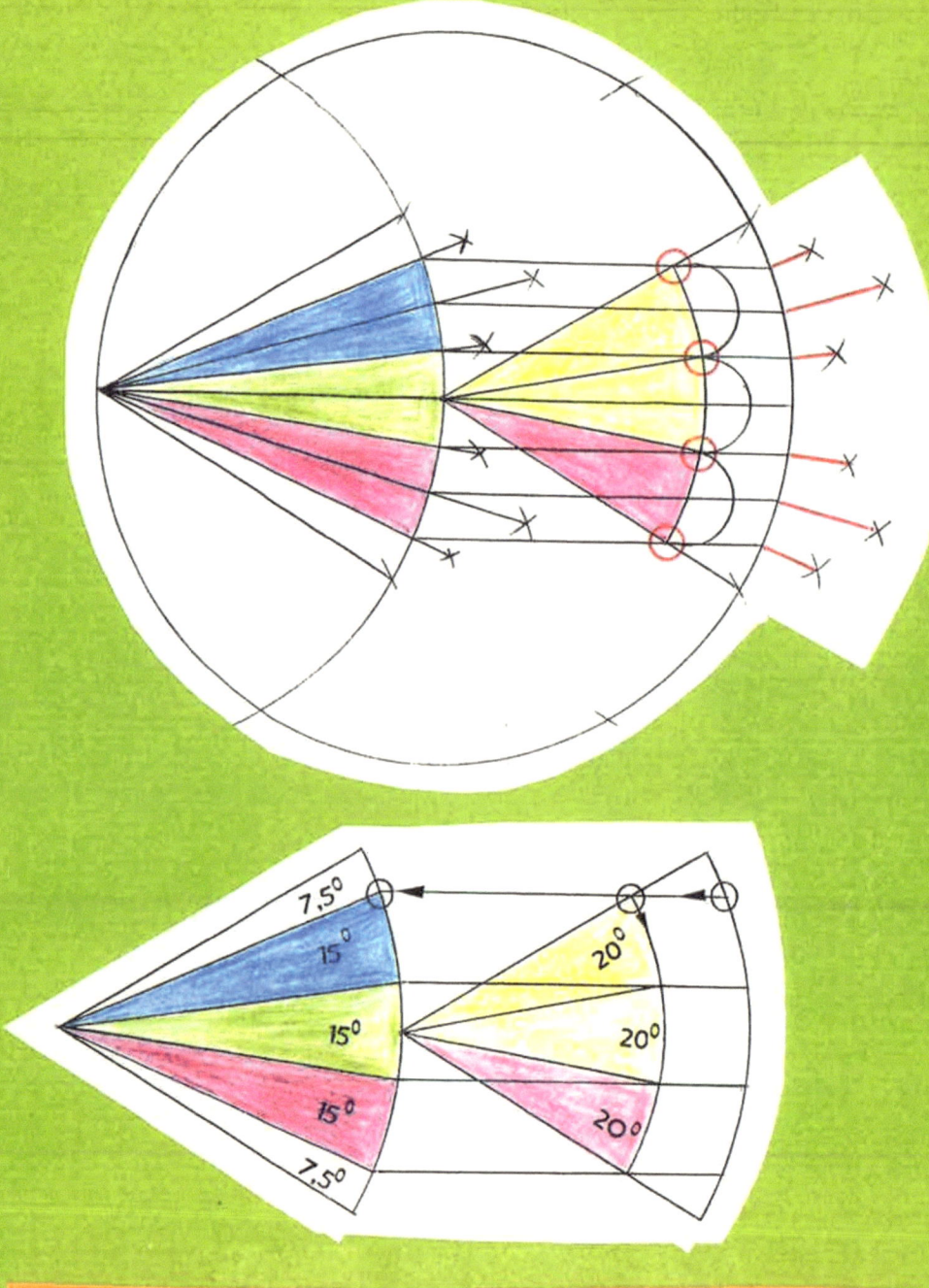

THE NEW GEOMETRY

Harald Schatz

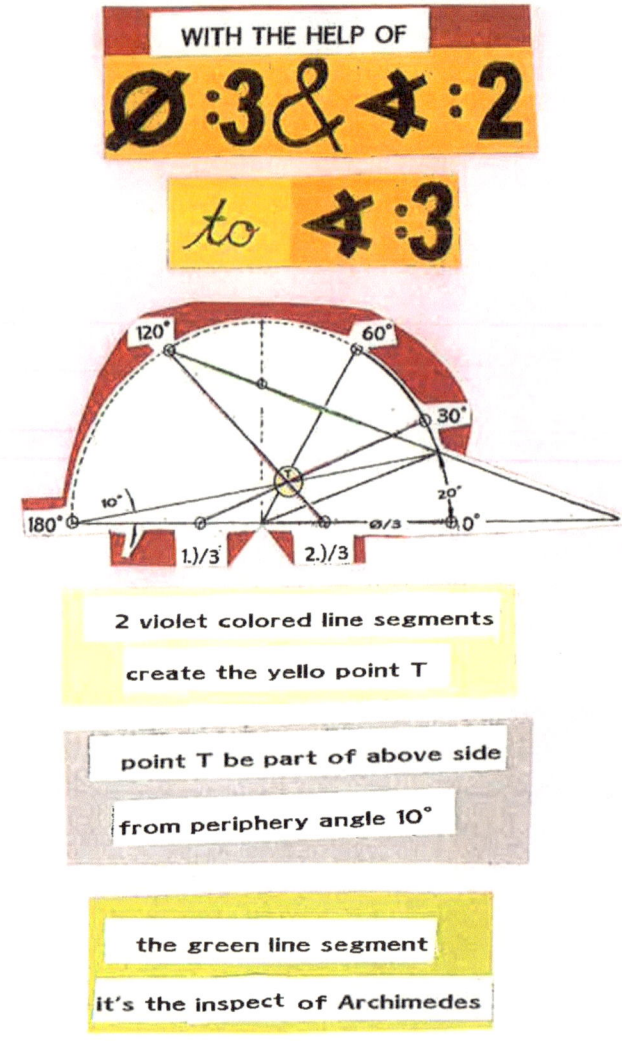

2 violet colored line segments

create the yello point T

point T be part of above side

from periphery angle 10°

the green line segment

it's the inspect of Archimedes

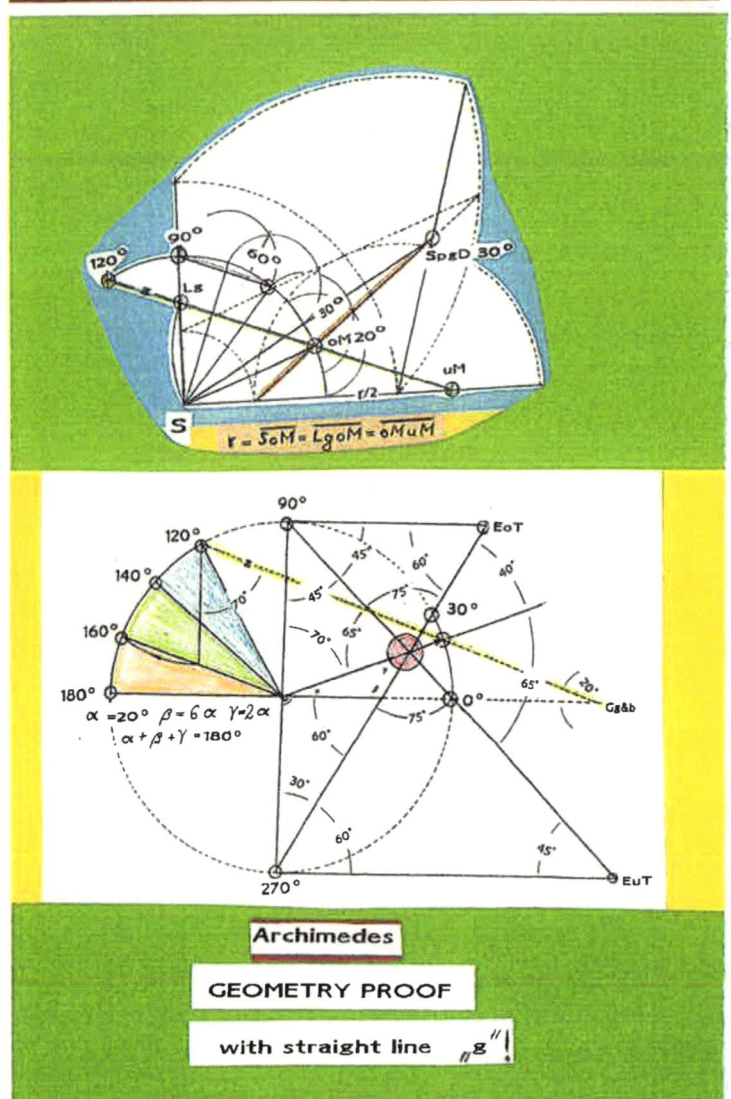

S $r = \overline{5oM} = \overline{LgoM} = \overline{oMuM}$

$\alpha = 20° \quad \beta = 6\alpha \quad \gamma = 2\alpha$
$\alpha + \beta + \gamma = 180°$

Archimedes

GEOMETRY PROOF

with straight line $_ng''$!

75° : 3

from 5   part of either stretch

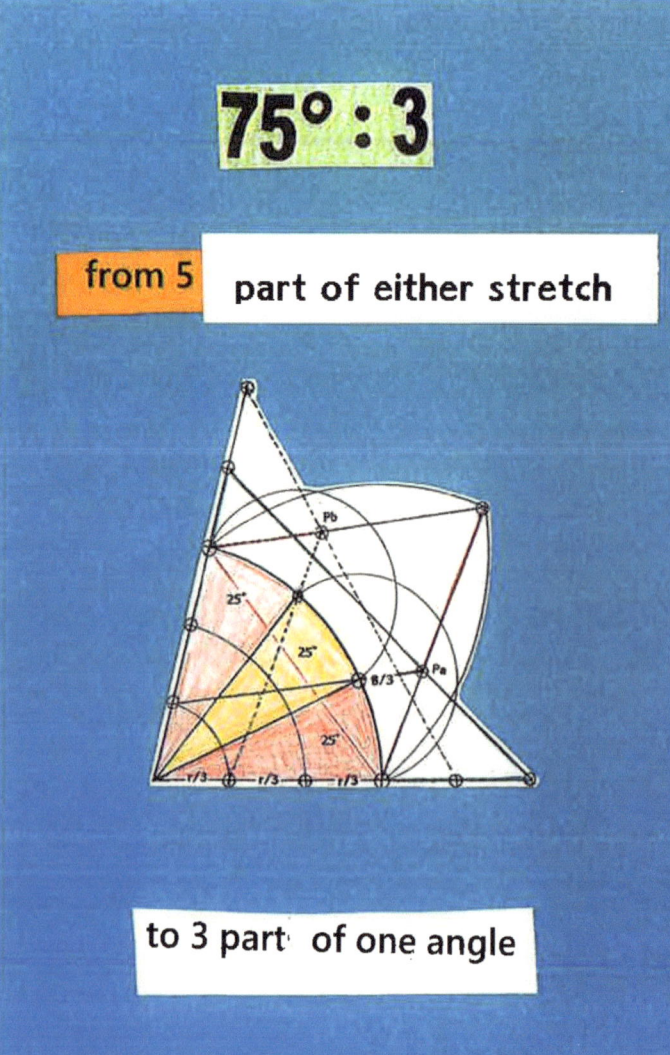

to 3 part of one angle

From 60°
to 20°

angle

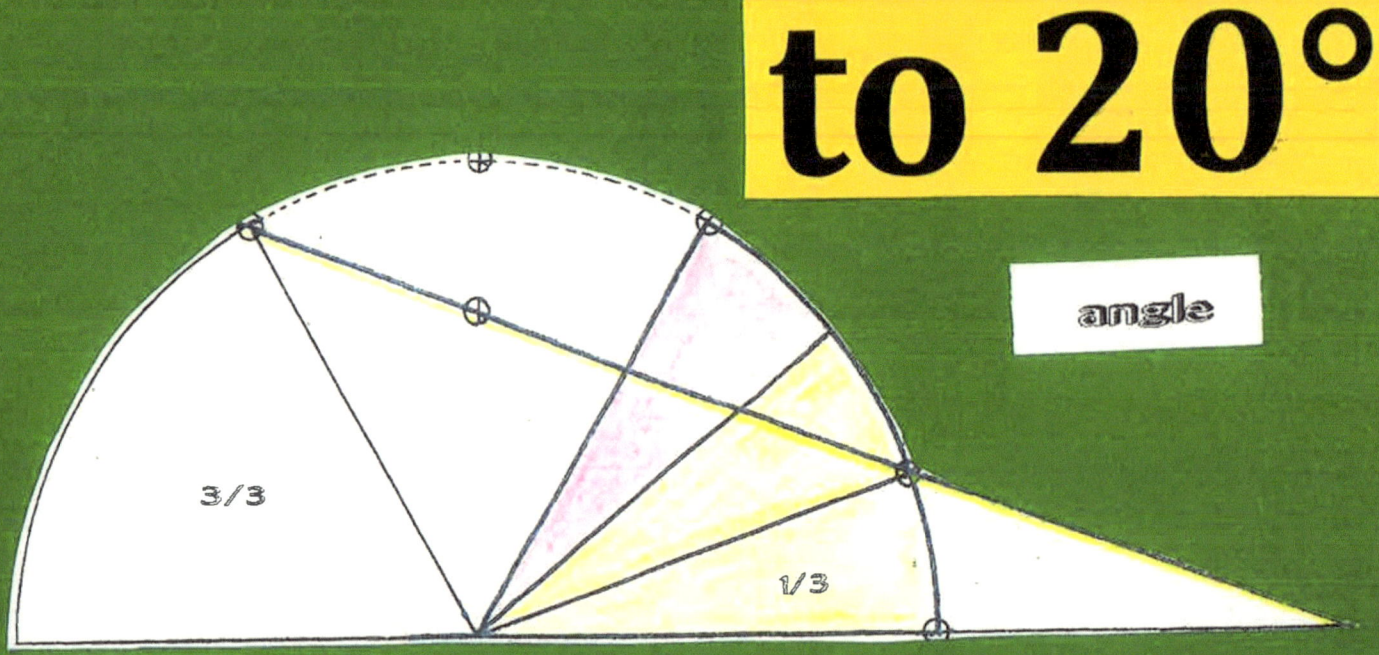

3/3

1/3

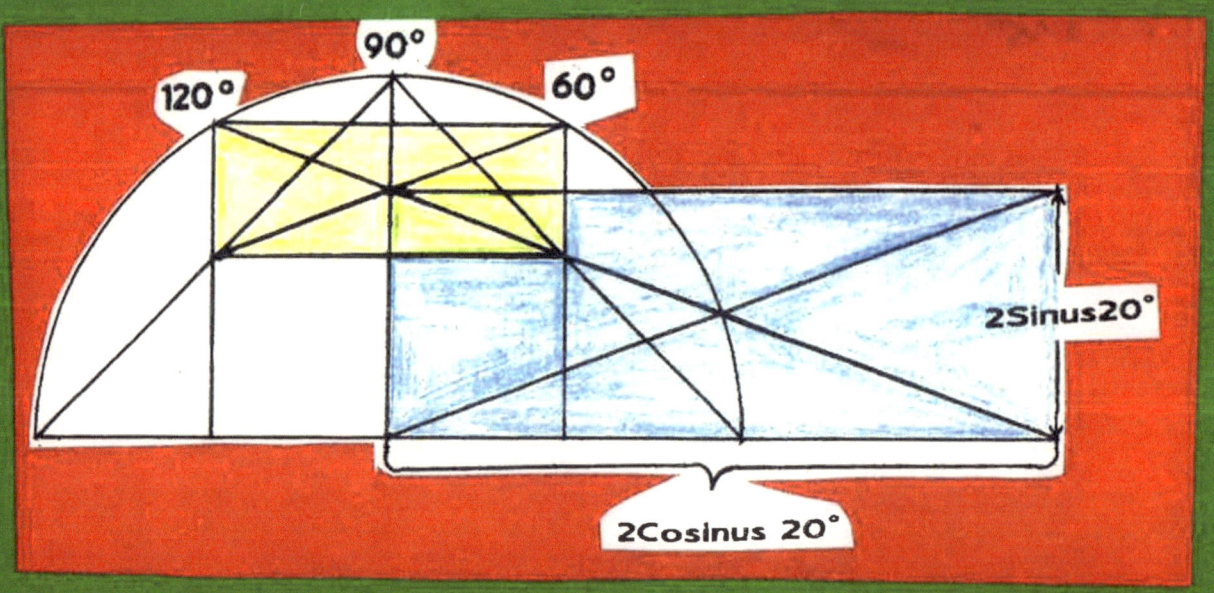

120° 90° 60°

2Sinus20°

2Cosinus 20°

with centre
points of rectangles

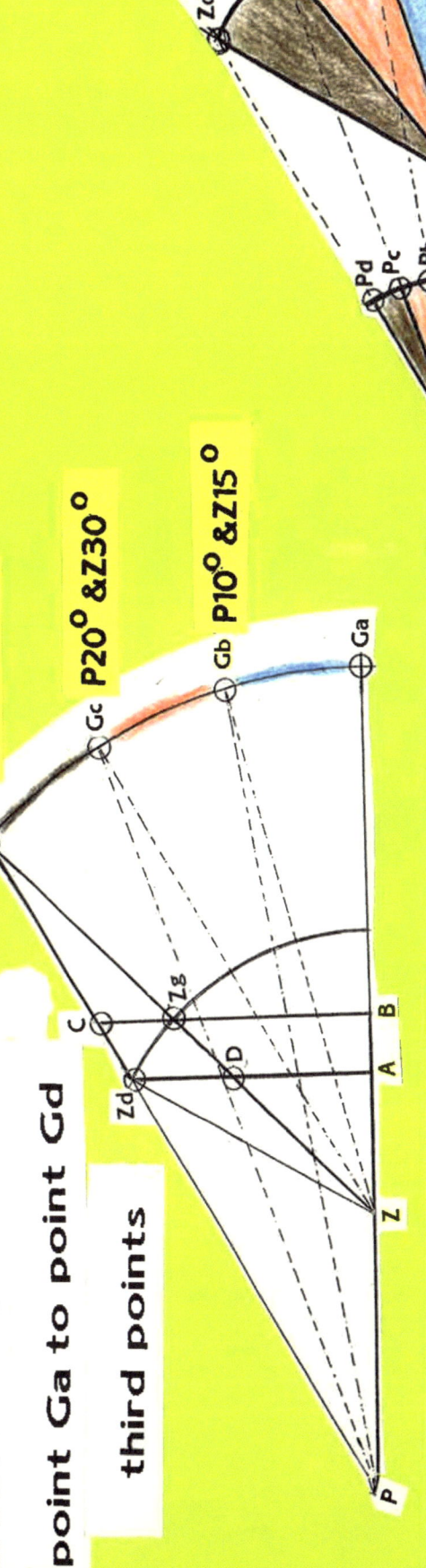

P " are points from periphery angle

Z " are points from centrical angle

$60° : 4 = 15°$

the stretches from point P and point Z

create the points from

point Ga to point Gd

third points

Gd P30° & Z45°

Gc P20° & Z30°

Gb P10° & Z15°

on arc from centrical angle and periphery angle

My address it`s

Harald Schatz
Käswasserstraße 58
90562 Kalchreuth

I am born on the 26.9.1949 in the
centre of Franconia Germany
Hersbruck
I wos a male nurse
the discovery it`s
commence
on 22.10. 2004

I make this Geometry
on July 2023 ready